Julian Liese

Germany - Southeast Asia

A scanty overview of the German maritime security policy and a short view on German perception and expectations concerning sea trade and other important maritime issues

GRIN Publishing

Bibliographic information published by the German National Library:

The German National Library lists this publication in the National Bibliography; detailed bibliographic data are available on the Internet at http://dnb.dnb.de .

Imprint:

Copyright © 2009 GRIN Verlag, Open Publishing GmbH
Print and binding: Books on Demand GmbH, Norderstedt Germany
ISBN: 978-3-640-92704-3

This book at GRIN:

http://www.grin.com/en/e-book/172658/germany-southeast-asia

GRIN - Your knowledge has value

Since its foundation in 1998, GRIN has specialized in publishing academic texts by students, college teachers and other academics as e-book and printed book. The website www.grin.com is an ideal platform for presenting term papers, final papers, scientific essays, dissertations and specialist books.

Visit us on the internet:

http://www.grin.com/

http://www.facebook.com/grincom

http://www.twitter.com/grin_com

Germany and Southeast Asia

- A scanty overview of the German maritime security policy and a short view on German perception and expectations concerning sea trade and other important maritime issues -

**S. RAJARATNAM SCHOOL
OF INTERNATIONAL STUDIES**
A Graduate School of Nanyang Technological University

Julian Liese, Lieutenant

Germany and Southeast Asia

- A scanty overview of the German maritime security policy and a short view on German perception and expectations concerning sea trade and other important maritime issues -

T his comment shall give an insight into the German perception with regard to its own maritime interests in the Southeast Asian area. First, the objectives of the government will be lighted up. Secondly, military, as well as economic aspects of the maritime domain in Southeast Asia will be discussed. The main focus will lie on the area of the energy production industry, the container transport and India as an ambitious maritime power.

I. G erman security interests within the scope of the European security strategy

T aken from the European security strategy as off 12/12/2003, stated in a position paper concerning the security strategy for Germany, by the German political party CDU/CSU of the Bundestag[1], the job of the German state is to preserve the values of the basic law according to right and freedom, democracy, security and well-being for the citizens of the country, to protect them against dangers and to protect the sovereignty and intactness of the German state as such. In addition duties and obligations which

---

1 European security strategy from the 12.12.2003: Security strategy for Germany
Decision of the CDU/CSU group of the Bundestag
The CDU/CSU group of the Bundestag has decided on a position paper for the security conference of the group.
The paper names Germany's interests within the scope of a European strategy and draws the conclusions from it.

arise from the membership of Germany in alliances like NATO and the EU have to be taken into account. Consequently in consideration of the called risks for our security it lies in our interest to prevent, if possible, or to contribute to master regional crises and conflicts which can affect our security and interests in the place of their origin. Global challenges like the menace by the transnational terrorism, the wide spreading of weapons of mass destruction or the results of the climate change have to be taken to contribute to the esteem of the human rights, for the propagation of freedom, democracy and rule of law (good governance), to promote free world trade including a secure energy supply and raw material care as a basis of our prosperity and to reduce the gap between poor and rich world regions on the basis of the model of the social market economy.

T he prosperity of the German society, as a main target for political action, is also part of the basic reasons for Germany's efforts in the maritime domain of Southeast Asia, because in the last decades the world-economic development was marked by a rapidly progressing intensification of world trade and the international division of labour. Caused by the expansionary development of international goods, oversea trade and maritime journeys were and are among the boom sectors during the past years and now. In particular, the increase of globalization's intensity, which occurred since the entry of the Republic of China to the general world trade agreement WTO in 2001, could hardly be foreseen[2].

T o be able to cope with the political challenges of a maritime security strategy for Southeast Asia and to be able to use their chances at the best, Germany needs a strategical action plan, not only unilateral, but also coordinated with the transatlantic partners of the EU. It is nessacary to strengthen the strains of the countries of asia for more democracy, political stability and development, a successful anti-terrorism warfare

---

2 Hamburgisches Welt Wirtschafts Institut
http://www.hwwi.org/

as well as to faciliate a stronger regional cooperation. Compared to the past, Germany must create a net of allies and like-minded people in the region with which common political aims are pursued, more strictly now.

G ermany's periphery changes. By the forming of more and more non-state actors; by the rise of new powers like India and China; by the growing importance and influence of NGOs. New needs arise from these changes and, beside the possibilities of cooperation dealing with solutions of global and regional security problems, which could come to life, also new conflicts can originate by power movements on the international level.

II. C onsequences for a German security policy

T he pursuit of its interests and strategic aims requires actions, that are more active, untimely, quick, coherent and robust, if necessary. Furthermore, these aims also need to be linked together more closely. This is valid for all instruments and abilities which are available in a crisis to cope and prevent a conflict.

H ome protection, as the oldest and still highest goal of the basic law also encloses aspects of maritime security, such as the protection against possible terrorist attacks, the security of German ships and their routs and traffic, for the environment and for fishing. For the maritime security beyond German territorial waters on the high seas and - with approval of the respective partner nation - if necessary also in foreign territorial waters unequivocal juridical conditions for a more robust application of the German navy must be created against piracy or terrorism.

III. E conomical reasons for a glance to the Southeast

D uring the last years the globalization has taken place with great intensity. Who invested in ships between 2001 and 2002 – now, rubs his hands. Huge profits have resulted. Maritime promotional capacities have become a scanty property. This encloses the harbour arrangements and hinterland bindings. A "fight for the quays" has broken out long ago. Long downtimes, but also increased security standards result in exploding costs.

H ow do the single limbs of the logistics chain, e.g. producers, forwarding agencies, harbour operators, shipping companies, but also state places or supranational organizations react to these challenges? And which chances open for investors?

F irst of all, the economic centre shifts to Southeast Asia.
S econd, the decisions on production locations, now depend on a functional transport infrastructure more and more, beside low labour costs or the degree of the education.
F urthermore ships remain, by far, the most energy-efficient and most eco friendly vessel to carry goods and people within the range of transportation means. Special ships will pull the biggest inquiry on themselves.

"T he progressive worldwide processes of integration [and] the [...] dismantling of commercial obstacles and the increase to be expected of the prosperity in numerous regions of the world will also bring a clear expansion of world trade with themselves in future and require an expansion of the freighter journey and harbour capacities", according to Prof. Dr. Thomas Straubhaar, manager of the Hamburg world economy institute (HWWI)[3].

---

[3] Hamburgisches Welt Wirtschafts Institut

T hen trade measured, in Euro, will expand with an average growth rate of suggested 6.6 percent during the next 24 years. At the same time the commercial volume will grow, in tones, merely about 3.3 percent p.a. in the same period, because increasingly the first (weight-reducing) refining steps take place in the raw material delivering countries themselves and, in addition, the trade of high-quality and relatively light industrial goods and consumer goods will increase between the regions themselves. The historical consideration of the distribution of goods on the different transport bearers (Modal Split) points to the fact that the traffic bearer's ship for the international exchange of commodities till 2030 might only change a little. Then the acceptance of a steady Modal Split implies that also the sea trade volume p.a. might expand about 3.3 percent.

B ecause trade with container goods will rise further in maritime transport, exceptionally harbours with a high container moving capacity will strongly profit from the general growth of the maritime transport. Besides, it is obvious, that in particular those harbours whose trans-shipment centres stand in connection with very expanding regions like, for example, Southeast Asia will profit and grow the most.

IV. S trategical reasons for a glance to the Southeast

T he trends named above go hand in hand with strategical draughts, which have their origins already in the 90s. On the one hand they are based on an enhanced focus on non-military and non-traditional security dimensions (i.e. drug smuggling, piracy, environmental destructions, illnesses / epidemics, increasing competition is also calculated around economic resources, and unequal poverty distribution) and, on the

---

http://www.hwwi.org/

other hand, rather domestic security challenges (secession, piracy, terrorism and others) within the security policy and defence policy of nation states.

These complex problems are based on an even more vague background of Islamic terror groups in Eastern Asia itself and other „non-state actors" which operate in and from "weak" or even "failed" states ("failed states").

To the potential and varied security challenges of the Asian-Pacific region, with special consideration of Southeast Asia, in particular the following factors have to be recognized[4]:

- Serious environmental problems
- Migration (i.e. refugee movements)
- Other non-traditional security challenges (like international criminal activity like drugs smuggling, people smuggling and smuggling of arms, environmental pollution and others)
- The absence of a wide regional security order for the whole Asian-Pacific region (institutionalizations of the Association of South-East Asian Nations are no proper web of boundaries, yet)
- An increasing power-political competition between ambitious China as potentially future hegemonial power and the USA, Japan and India, which applies increasingly to South-East Asia and his geopolitical structure (above all concerning the navigation ways) since the middle of the nineties.

Hence, the "Sea Lanes of Communication" (SLOCs) in Southeast Asia and in the South-Chinese Sea have a central strategic significance for all bordering states. About 75 percent of the Japanese, South Korean and Taiwanese oil imports, more than 30 percent of world trade and 80 percent of Asian-Pacific trade occur about the SLOCs in the South-

---

4 Bayerische Landeszentrale für politische Bildungsarbeit: Bayerische Zeitschrift für Politik und Geschichte
http://www.km.bayern.de/

Chinese Sea. In addition, about two thirds of the transport of shipped liquefied natural gas goes through the South-Chinese sea. Above all, special strategic significance has the bottleneck (Choke-Point) of the Malacca street which is 630 km long and is just 1.5 miles wide at the most narrow point, as well as the harbour of Singapore as one of the worlds biggest commercial trading centres[5]:

- More than 50,000 ships per year and 140 per day steam through this potential focus where Islamic fundamentalism, piracy, poverty, corruption and nepotism meet in Southeast Asia.
- One third of world trade and 50 percent of the global oil transport take place there.
- Till 2005 up to 40 percent of the worldwide piracy raids (more than 300 attacks worldwide with more than 30 dead people) were to be stated.In comparison only 93 registered attacks took place in the year before in Indonesian waters, followed by 37 in the Malacca street, in spite of the common sea patrols of Indonesia, Singapore and Malaysia.
- At present the closing of the harbour of Singapore yearly would cost more than 200 billion US dollars p.a..

V. S ea lanes, energy, trade, piracy and terrorism

N umerous regional experts no longer neglect the possibility of a growing cooperation between traditional piracy and international terrorism in the future.

This could entail a maritime terrorism and with it substantially more dangerous attacks on big ships (i.e. oil tanker). Like the USA, Germany has to develop new regional and global security initiatives (as the regional Maritime Security Initiative/ „RMSI" or the

---

5 Bayerische Landeszentrale für politische Bildungsarbeit: Bayerische Zeitschrift für Politik und Geschichte
http://www.km.bayern.de/

Proliferation Security Initiative/ „PSI") within the frame of NATO and EU, to be able to guarantee the security of the international navigation also in the future.

F urthermore an increase of energy imports of the Southeast Asian region, accompanied by a continuous increase of trade with other states and regions, the ship traffic is expected to triple up to 2010 in the South-Chinese Sea.

Beside environmental issues this growing maritime trade throws up numerous safety-political questions (piracy and mine warfare for example).

The gigantic power demand of China depends on the same energy resources and maritime transport's routes like that of the USA, Japan, India and other industrialized countries as well as the developing economies in the industrial threshold countries, above all in Southeast Asia.

G ermany and the EU ignored most of the effects of the rapidly rising Chinese power demand on the global stable energy supply and the world order. Its strategical and global political significance were misjudged or mostly overlooked within Europe for a long time.

However, the energy and resources hunger as well as the dynamic economic development and its energy-political consequences for the foreign affairs and security policy of the regional states and the regional and global political stability have to be considered as a main topic in Germany's foreign affairs.

A lso important is the construction of Chinese naval bases (e.g. in Pakistan) and Secret Service facilities along the Indian ocean, the Malacca street and the South-Chinese Sea, as a component of a long-term security strategy for the rapidly increasing maritime energy import from the Middle East (above all Iran and Saudi Arabia) and a naval strategy including investments for 30 to 40 years to built up the "Blue Water Navy", which then shall be competitive with the USA. This has safety-political challenges not only for the USA, India, Japan and the Association's of South-East Asian Nations states, but also for Germany and the EU.

VI. China, India and the problem of upcoming superpowers

The presence of a – friendly – hegemony power in the Southeast Asian region, the security and stability it creates without restricting its satellite states, is qanything but uninteresting, not only for the USA. China is no possible aspirant for this job. Also not from Germany's point of view. It lost all credibility by the exorbitant possession claims in the South-Chinese Sea and the attempt to enforce these claims by military suppression. The "hyper power" USA, which, since the 11th of September, 2001, is increasingly involved again in the region, is seen with mixed feelings and rather liked to be multilaterally integrated. But India however, economically active in South-East Asia, but not dominating, without virulent conflicts with ASEAN states and beside that, also a democratic constitutional state, appears as an inspiring confidence cooperation partner. Though the Indian armed forces will be strong enough to promise protection after the ongoing modernisation, it will by far not be strong enough, to intimidate[6].

India's military already adapts itself to the changed strategical requirements. After the whole Indian ocean had been defined routes as an Indian security zone, including the adjoining seas, the first Indian naval strategy of 2004 specified the necessary capacities of the Indian navy: Ability to remote power projection far from the own coasts, sea supervision in own and foreign waters and concentrated attacks against hostile facilities.

Besides, India also has an advantage within the modernization of its weapon systems towards China, which struggles because of sanctions for every technology transfer for its

---

[6] Möller, Kay: Maritime Sicherheit und die Suche nach politischem Einfluss in Südostasien. SWP-Studie, Berlin 2006

armed forces and build up its outdated military-industrial complex with own resources. Foreign weapons almost exclusively come from Russia – and this hesitantly.

I n comparison, almost the whole range of articles of the international weapon industry is available for India's armed forces, which, by the way, is the third-biggest of the world. France and Germany deliver submarines of the brands Scorpion and Type 209, the USA airplanes, helicopters and landing support ships for amphibious operations, Israel technical equipment. Furthermore India signed a contract concerning the Phalcon-radar systems (Israel), which are needed for the construction of a aerial-supported supervision net, similar to AWACS. China had failed before with a similar order in Israel because of an US veto. Even the Russian federation prefers India as a trading partner with such sensitive goods, because it must not fear that its weapons are used on a common border against themselves one day. So there is no need to wonder that the thickest fish, which the Indian navy could pull on board, comes from Russia: The Russian aircraft carrier Admiral Gorshkov, which is renamed to "Vikramaditya". The carrier offers quite new possibilities to India's naval pilots[7].

T he oldest and most important partner of India, within the Southeast Asian region, is Vietnam, with which already safety-political relations exist since the Vietnam War. They were deepened in March, 2000 to a strategical partnership in numerous political fields. The cooperation encloses Indian supply of arms and servicing of Vietnamese weapon systems from Russian production as well as cooperation by the sea supervision in the sensitive South-Chinese Sea. In February, 2006 Delhi signed an agreement about military cooperation with the Philippines, which contains the education of military professional forces through India. Australia agreed, not only to cooperation within the terrorist and pirate fight, but also gave to understand that even the delivery of Uranium is

---

7 Deutsche Bank Research: Indien – Auf dem Weg zur Weltmacht? 8. February 2006

conceivable for nuclear industry of India in July, 2007. One of the latest agreements was signed in October, 2007 with Singapore with which India already holds common naval exercises since 1993. It intends a cooperation of the aerial armed forces of both states what allows India to take a look at the modern weapon systems of Singapore's Air Force (RSAF), while Singapore can evacuate some of its weapon exercises from the own congested airspace to Indian air bases.

L ast the operation Sea Wave, the big relief operation of India for the areas affected by the Tsunami in Southeast Asia, Sri Lanka and the Maldives in the beginning of 2005, must be also seen in this context. At that time 32 ships of the Indian navy carried out the biggest non-military operation of its history – a clear signal to the eastern neighbours.

It is an important matter to India itself, as well as for Germany, Europe and also America. Therefore it is to be directed advisable for the German foreign affairs to deal with this fact[8].

---

8 http://www.suedasien.info/analysen/2311